HOW TO IMPROVE YOUR TOTAL SCORE ON THE SAT

THE SECRET CONFESSIONS OF AN EXPERT TUTOR

TIM CHAMPION

PLOTWORKS PUBLISHING

How to Improve Your Total Score on the SAT

The Secret Confessions of an Expert Tutor

By Tim Champion

 The author is unaffiliated with Educational Testing Service or the College Board. He is, however, very affiliated with a major test-prep company whose name you would recognize.

Electronic ISBN: 978-1-960936-55-4

Print ISBN: 978-1-960936-56-1

1

How To Improve Your Total Score on the SAT is not a textbook. It's not full of fourteen thousand practice questions. There are innumerable other books that offer those things.

This is a *handbook*.

Handbooks have a long and glorious publishing history. They have gone by many names—how-tos, manuals, pocket references, almanacs—and they teach many things, from how to fix your car to how to tell the infidel that his body is about to die but that his soul will be spared by a just and merciful God. My favorite handbook, *The Art of Courtly Love*, by Andreas Capellanus, taught young strolling minstrels in medieval Provence how to write irresistible love songs that would lure the prettiest milkmaids in the parish.

This handbook has one purpose:

It will show you how to think like a bastard.

That's not an overstatement. The SAT, which stands for nothing (really, look it up), is manufactured by ETS, which also stands for nothing (spiritually, at least). In real life, ETS actually stands for Evil Testing Service, and this is an organization full of bastards. *Bastards* might even be too soft a word. What do you call people who lie without compunction—and then pretend like they *didn't* lie when you catch them?

I believe that word is *sociopath.*

They pretend to be sympathetic and fair. That's not true. Only a sociopath would hand you a cup of tea while plunging a knife into your back. I'm halfway convinced that the test makers torture kittens in their basements.

I'm exaggerating, of course, but not by much. ETS actually stands for Educational Testing Service. They make every standardized test known to man—the entrance exams for the CIA, for beauty school, for golf pros. After you die, you even have to take a standardized test to get into heaven.

Kidding about that last one. But the CIA, beauty school, and golf pro exams? Those really are ETS too.

Still, don't forget: *They are bastards.*

This handbook is going to be a longish/shortish piece of writing—too long for most magazines, too short to occupy space at your local nonexistent bookstore—and it's going to show you how to raise your score on the exam.

To do that, you need to learn to think like a bastard. To read like a bastard. To write like a bastard. If you've done any acting, it should be easy. It's a lot like slipping into character.

Let's learn the role.

MATH

2

The first thing to know is that standardized test math isn't *real* math. It's a weird beast, a strange hybrid of mathematics and English. It's like *verbal numerical problem-solving*.

Funny story: I once taught with a really smart guy who had finished a Ph.D. in math. He'd started tutoring the SAT after he'd graduated, while he was applying for jobs as a professor. I asked him what he thought of standardized testing math problems, and his face turned red. "It's not *real* math!" he said, sputtering with rage. "There are too many ... *words*!"

That, in a nutshell, is the problem—or the advantage, depending on how you look at it. The bastards who make the SAT aren't testing you on real math. In some instances, they're using mathematical concepts as a way to see how closely you can read. In other words, they're trying to trip you up with sneakily phrased sentences. That means that the syntax matters.

There are two math modules on the digital SAT. In each, you are given 35 minutes to complete 22 questions. So that's a total of 44 questions in 70 minutes.

The first module contains a mix of easy, medium, and hard questions. However, the adaptive nature of the new digital SAT means that your second math module will either grow more difficult or become easier—depending on how you performed on the first one.

Nervous yet?

ETS has been using this model for a decade already with a different exam, the GRE, which used to be required by a lot of graduate school programs. Now the bastards have basically copied the same format to the high school exam. I'm surprised it took them so long to do it. I expected a digital SAT back in 2015, but it never materialized.

You have full access to an onscreen Desmos calculator, whenever you need it, which is built into the Bluebook testing app. If you don't want to use that calculator, you can bring your own to the testing site. There are limits to the types that you're permitted to bring, however.

There are two question formats: multiple choice and "student-produced responses". Everybody knows multiple choice, and 75% of the questions in the math modules consist of this type.

Another 25% of the questions, however, will be student-produced responses, which are better known as fill-ins. In other words, you type the answer. The exam allows you 5 characters for a positive answer, and 6 characters for a negative answer. If your answer is a mixed fraction, like three and a half, please don't write 31/2, because that looks like a very different answer to the scoring software. Instead, convert to an improper fraction, e.g. 7/2.

You never need to enter symbols like percent signs, commas, or dollar signs.

Sometimes this type of question causes some students to self-destruct. Some students lose the psychological crutch of seeing those four reassuring answer choices. Don't let that be you.

Lastly, two questions in each module are experimental —that's four questions total—and don't count towards your score. You won't know which ones are experimental, and don't try to guess which ones they are either. Treat them all the same.

HERE's something else about scoring.

If you're reading this handbook, it's possible that you're not very mathematically inclined. That's fine. I've literally helped a thousand students like you, some of whom were totally paralyzed by low self-esteem as soon as they so much as thought about an addition problem. Don't worry about that. I would urge you to remember, when you feel like you're not good enough, that we all can't be good at everything—and colleges know this better than anybody. They don't expect perfection from you, no matter how much your parents and counselors may have tried to brainwash you.

In fact, if you're planning to go into the liberal arts, here's a radical piece of advice: Just show colleges that you're *decent* with numbers.

Here's an example. I once had a student—let's call her Maria—who was an incredibly verbal young lady, always reading, always writing, a real wordsmith. She'll probably win a Pulitzer Prize someday, or maybe edit a dictionary. In fact, her reading and writing scores were around 750.

However, all her life she'd harbored a serious distaste for numbers, and her math score had been hovering in the upper 400s.

The problem: She wanted to go to Barnard College, a women's college in New York that's affiliated with Columbia University. When the admissions department there took an early look at her SAT score, they said one thing:

We can't admit you until you score a 550 on the math.

Keep in mind that they wanted her. Badly. Barnard was basically saying *Please don't make us reject you.*

So her family called me in to help. For eight weeks, Maria struggled and cried, weeped and moaned, gnashed her teeth and chucked her fancy bookmarks across her dining room table. But with my help she eventually limped across the finish line. She scored a 570. Today she's graduated from Barnard and has become a respected young journalist. (I helped edit her application essays too.)

There are two lessons here. One, I'm very good at tutoring math. Two, if you know you're going to be a liberal arts major, don't worry about getting a perfect score, or even scoring a 700. It's not necessary. College admissions committees don't punish you for what you're lacking—they celebrate you for what you *are*. Show excellence in other stuff, do well enough on math, and more often than not, they'll show mercy.

For those who don't care much for math: I recommend scoring a 600.

That should be enough for most universities, if you're a word nerd, since the average score on math is about 500, and a standard deviation is 100 points. One standard deviation above average should satisfy the vast majority of college admissions committees out there, particularly if you inform them in your application essay that you're going to study the liberal arts.

And the best part of this deal? Scoring a 600 on the math isn't *that* hard.

Someone once said that it's wise to learn from your mistakes, but it's even wiser to learn from the mistakes of others. In my career as a tutor of standardized test math, I've watched people commit the same mistakes, stumble into the same tiger pits, over and over. Through long experience, I have become able to predict the mistakes you are going to make long before you make them.

This handbook is a list of those errors, explained in patient detail. I want you to learn from the mistakes of others.

It begins in the next chapter.

3

Rule #1: You almost never need a calculator for fractions, exponents, or roots.

HAVE you ever been confronted by a complicated fraction question? Did it scare the Uggs off your feet when you remembered that you haven't worked with fractions since middle school? Did you then try to translate all the fractions into decimals using your calculator, even though you knew that the multiple-choice answers were all in fractions, and that you would have to change everything back?

This is possibly the most common mistake I see students making. It's much, much easier to work with fractions on paper than it is to convert them into decimals and enter them into your calculator. Remember that time is of the essence on this test.

How do you solve fractions? We all knew how in the seventh grade, and we've all forgotten. The good news is that it hasn't changed: *To multiply fractions, multiply across the top and across the bottom.*

Voila, you're finished.

Dividing fractions is the same, but with one more step. *To divide fractions, just flip one of them upside down, then multiply across the top and across the bottom.*

Adding and subtracting fractions can be a bit trickier. Just change the bottom number (the denominator) to the lowest common number that both have in common. This is called the LCD (lowest common denominator). You can find that by choosing one of the numbers and multiplying it by 2, 3, 4, etc until you arrive at a number that the other denominator also goes into. Then multiply each numerator by whatever factor (2, 3, 4, etc) that you multiplied its denominator by. Add or subtract across the top and keep the denominator the same.

Exponents are another part of arithmetic that send students scurrying for their pink TI-30Xs. Don't bother. It's faster and easier to do them on paper, and the SAT usually is testing you on whether you remember how to do it that way.

Don't panic. The world has created a handy mnemonic device to help you remember how to deal with exponents.

It's called MADSPM. Here's what it means.

If you MULTIPLY the exponents—

—then ADD them. $5^2 \times 5^3 = 5^5$

If you DIVIDE the exponents—

—then SUBTRACT them. $5^5/5^3 = 5^2$

If you POWER the exponents to one another—

—then MULTIPLY them. $(5^2)^3 = 5^6$

. . .

AGAIN, MADSPM.

Keep this simple rule in mind, and you'll be able to solve basic exponent questions in about twenty seconds. That's no exaggeration.

It's slightly more difficult when you see negative or fractional exponents. As a reminder, the basic rules here:

- For a negative exponent, turn the exponent positive, put it at the bottom of a fraction, and put a 1 in the numerator.

$$x^{-2} = \frac{1}{x^2}$$

- A fractional exponent is even easier. Just turn it into a root. The numerator of the exponent becomes the new exponent, while the denominator becomes the type of root.

$$x^{\frac{2}{3}} = \sqrt[3]{x^2}$$

ANOTHER TYPE of question that strikes fear into a typical math-averse heart is how to combine roots.

Mostly because we've forgotten how to do them without a calculator. Trust me when I say that, on the SAT, you pretty much will never need to use a calculator with a root. Like exponents, doing them on paper is easy.

All you need to remember is the *Prudes and Sluts* rule.

Yeah, you read that right.

Prudes and Sluts.

That's a cute shorthand to know that when you *add* or *subtract* square roots, they are *prudes*. In other words, they only mix with other roots that look just like them. For example:

$$3\sqrt{6} + 4\sqrt{6} = 7\sqrt{6}$$

A $\sqrt{6}$ will only add or subtract with another $\sqrt{6}$. Here's another metaphor: think of a square root as a pineapple. In the question above, 3 pineapples plus 4 pineapples equals 7 pineapples. To do this, however, the number inside the square roots must be the same. They must all be pineapples.

Prudes.

What about this—

$$3\sqrt{6} + 3\sqrt{7} = ???$$

Trick question! There is no answer. It's impossible to solve this any further. The roots won't budge. Their mother told them that nice girls don't go out with bad roots. You can't add pineapples to mangos. They don't even add together in smoothies very well. There are a hundred metaphors to describe what's happening here.

On the other hand, when you *multiply* or *divide* square roots, they become very slutty. Very promiscuous. They will mix with all other roots regardless of what number may be inside. They are the Kardashians of math. Every night is a drunken hookup. Look:

$$3\sqrt{6}\,(3\sqrt{7}) = 9\sqrt{42}$$

These numbers wouldn't even glance at each other in an addition or subtraction problem. But here, in the basement nightclub meat market of multiplication and division, they're jamming tongues down each other's throats.

How do you multiply or divide them? Just multiply the outside numbers together, then multiply the inside ones together. Then factor the square root, if necessary, which you will usually have to do to get one of the answer choices. I presume you remember how to do that.

That's it. Division works the same way. Divide the outside numbers, divide the inside numbers.

Review all these principles, especially if you're missing a lot of easy- or medium-level questions.

4

Rule #2: Whenever you see variables in the answer choices, don't use algebra. Instead, make up a number and solve it like an arithmetic problem.

This is the most important technique that you need to raise your score. It goes by many names and has many disguises in many different test prep companies.

It can be summed up in three words:

Don't do algebra.

Even if you're good at algebra, *don't do algebra.*

I'll go further than that. *Especially* if you're good at algebra—*don't do algebra.*

Look at this problem:

If x and y are integers and xy is an even integer, which of the following must be an odd integer?

a) $xy + 5$
b) $x + y$
c) xy
d) $4x$

ANYONE WHO CAN SOLVE that problem using algebra—within the time constraints of the exam, of course—doesn't need to read this handbook. You're a future math professor and you don't need any help from the rest of us.

Most students, however, will struggle to write an equation, and will fail horribly, wasting precious minutes in the process. That's unnecessary.

Here's what you do instead:

1. Invent a value for the variable. It can be anything. 2, 3, and 5 work great. Stay away from 0 because it's the grim reaper of math and kills with its frosty breath anything that it multiplies or divides. Also stay away from 1 because it's like kissing your sister or eating a rice cake. It doesn't change anything in the problem, so why even bother?
2. Write down your made-up value. I'll make up these: x=2, y=4.
3. Then solve the problem like an arithmetic problem.
4. Finally, go to the answers. Write down your arithmetic answer for each one. My answers for the question above would be as follows: a) 13, b) 6, c) 8, d) 8.
5. Answer is A. It's the only odd integer.

This technique is an algebra killer, which is a good thing. The majority of test takers are better off for using it.

On other questions, you will find an arithmetic "answer", using your imaginary made-up values. I call this the *target number.*

For example:

$$A = \{x + 3, x - 4, 3x + 4\}$$

If $x > 0$, what is the median of set A shown above?

a) 2x-4
b) x+3
c) 3x+4
d) 3x+1

You SEE the variables in the answers? Hallelujah! We know how to solve this question:

1. Invent a number for the variable.
2. Write it down. x = 2.
3. Solve the problem. Set A = {5, -2, 10). We put them in numerical order and get A = {-2, 5, 10}.
4. What is the question asking for? The median, which is 5. That is your *target number*. Write it down and circle it three times, like a target.
5. Go to the answer choices. Plug in x = 2. Write down the answers you calculate: a) 0, b) 5, c) 10, d) 7.
6. The only one that matches your target number is B. So that's the answer.

Practice using this technique. Remember, *whenever you see variables in the answer choices, don't use algebra.* **Make up a number instead, and solve it like an arithmetic problem.**

In my experience, about thirty percent of the math questions use this technique, and they are drawn from every area of math that you can imagine. There are variables in the answers in geometry questions, ratios and averages questions, algebra 1 and 2 questions, and functions and graphs questions.

RELATED: Any question that uses the words *how much, how many*, or *what is the value* has an easy solution:

Plug in the answers.

That means check the four answer choices.

For example—

If $p(x)=^3$, the function p is defined by the given equation. For what value of x does $p(x)=16$?

a) 1/4
b) 2
c) 4
d) 16

Go to each answer choice and plug in that value for x. Find the one that gives you 16 on the left side of the equation. If you'd like, you can start with either B or C, and move up or down from there.

The correct answer is C.

Again, you could do algebra. For many test takers, however, it's easier to convert it to arithmetic by using the answers.

5

Rule #3: Read every single flipping word in the question.

Pardon the language, but I felt an acidic little burp of frustration rise up out of my very soul when I wrote that. I've watched countless test takers point a proverbial gun at their feet and pull the trigger—by failing to do one thing.

What's that thing?

They don't notice every word in the question.

Here's what I mean. Read this question and answer it—

What is the least of 4 consecutive even integers if the sum of these integers is 92?

a) 20
b) 21
c) 22
d) 23

Be honest. Did you notice the word *even*?

For reasons known only to God, most people don't see the words *even* or *odd*, and miss this type of problem. I wish I could explain this phenomenon more precisely, but all I can say is that it tends to be lower-case adjectives that evade our reading comprehension. The bastards who make this test know this weird part of our human mental weakness. They exploit it ruthlessly.

To avoid this trap, read the sentence out loud (on your lips) with the tip of your pencil dragging beneath each word. It's strange, but it works.

To solve this particular question, you have two options: algebra or plugging in the answers. I recommend plugging in the answers. If you start with A, you'll quickly see that $20 + 22 + 24 + 26 = 92$.

The correct answer is A.

6

Rule #4: Break up word problems into bite-sized pieces.

THE SAT KNOWS that word problems are the bane of our existence, which is why there are so many of them.

I know that it can be intimidating to read some of the longer questions. Remember, you don't have to work on every question, especially if you are aiming for a 600. If you see a question written more complicatedly than a Nathaniel Hawthorne paragraph, feel free to skip it and come back later, no matter where in the module you encounter it.

When the time comes for you to work on those questions, remember to put your head down and take one step at a time. It's like climbing a mountain. You never look at the peak, because you'll just start crying and then sit down on a rock to eat all your trail mix before you've even started the hike.

Here's an example:

Rowena bought four rare LPs that cost $115, $105, $70, and $60, respectively. If she paid 1/5 of the total cost immediately, and the remainder in four equal installments, how much was each of the four equal payments?

a) $40
b) $55
c) $70
d) $100

Keep calm. Do one step at a time.

1. Add up the cost of the LPs. Total: $350.
2. Multiply by 1/5. Result: $70.
3. Subtract 70 from 350. Result: $280.
4. Divide 280 by 4. Answer: $70. Answer is C.

That's all. This is mostly an exercise in managing your own anxiety. Students who miss these questions often (but not always) have anxiety problems, sleeping problems, etc. Those types of personal issues exceed the scope of this handbook, but I will say that if you feel yourself panicking, just move along, and come back to this type of question when you've calmed your mind.

7

Rule #5: Know the difference between factors, prime factors, and multiples.

THE SAT MATH tests you on vocabulary in subtle ways. One of those is the difference between factors and prime factors.

The definition of a factor is easy: it's any two numbers that multiply to make another number.

To solve, make two columns of numbers that multiply to the number in question. Start with 1 in the left column, and the actual number in the right column. Then raise the number in the left column to the next factor, and match with the appropriate factor on the right.

WHAT ARE the factors of 24?

1, 24
2, 12

3, 8
4, 6

Keep making those columns until the two numbers meet. Then stop. It's like two elevators meeting halfway up a building—one going up, the other coming down.

For prime factors, you have to become a gardener. That's an incredibly lame joke that will only become apparent once you read the next sentence.

You should make a prime factor tree.

First be sure that you remember the prime numbers, so let's review those first. Prime numbers are defined as any number that is only multiplied by 1 and itself. You might think that the first prime number is 1.

It's not.

The first prime number is 2. Apparently, mathematicians have long, boring, abstract arguments about whether or not 1 is a prime number. The SAT has sided with those who say that it's not, so we'll just nod our heads and agree.

The official list of prime numbers up to 20 goes as follows:

2, 3, 5, 7, 11, 13, 17, 19

There are more, obviously, but you really don't need to know them for this test. Meanwhile, here's how to do a prime factor tree using the number 24.

24
/\
2 12

/\
2 6
/\
2 3

When you've finished it, circle all the terminal branches. In this case, 2, 2, 2, and 3. Those are your prime factors.

Factors and prime factors are always smaller than or equal to the number being discussed. Multiples, on the other hand, are larger than the number.

The multiples of 24, for example—

24, 48, 72, 96, 120 ... and so on.

Note: The first multiple of any number is itself. In other words, the first multiple of 24 is 24. The second multiple of 24 is 48. I've facepalmed myself a hundred times watching students go down in flames because they forgot this simple rule and believed that the first multiple of 24 was 48. That's totally wrong, and the bastards know this error, and so they lay this stupid little trap all the time, knowing it will snatch points out of the jaws of at least a few hungry test takers.

Sadists!

8

Rule #6: Memorize the groups formula.

STUDENTS RARELY ENCOUNTER the groups formula because almost nobody teaches it—not teachers, not tutors, not test prep companies. It's mind-bogglingly easy, and it's an automatic point.

Here's the groups formula:

TOTAL = GROUP 1 + GROUP 2 – BOTH + NEITHER

Memorize that. Tattoo it on your left forearm.

It's used in those types of word problems that involve two groups of people undertaking some activity, and a few people are shared by both groups, and a few people aren't in either.

For example:

In a class of 30 students, 11 students play violin, 14 students play piano, and 8 students play neither violin nor piano. How many students play both violin and piano?

This might send lesser test takers into paroxysms of panic, but not you. Just write the groups formula, and then fill in what you know:

Total = Group 1 + Group 2 – Both + Neither

$$30 = 11 + 14 - B + 8$$
$$30 = 33 - B$$
$$-3 = -B$$
$$B = 3$$

That's all there is to it. It's just a Venn diagram written algebraically. I'm baffled why school districts don't reveal this little trick more often.

9

Rule #7: Memorize the Special Right Triangles and Pythagorean Triples

ALTHOUGH THE SAT gives you a whole page of geometry formulas at the beginning of every math module, that doesn't mean that you should rely upon it. This is particularly true if you're a slow test taker.

The special right triangles are by far the most important concept in geometry to know. That's why you should *memorize* them.

$$45\text{-}45\text{-}90 \text{ TRIANGLE} = x, x, x\sqrt{2}$$
$$30\text{-}60\text{-}90 \text{ TRIANGLE} = x, x\sqrt{3}, 2x$$

Those lengths represent the side-side-hypotenuse of each right triangle.

Another thing to remember about them is that each of those two triangles is found within other shapes.

- A 45-45-90 triangle is half of a square, so at times you'll need to slice a square in half diagonally, and then calculate the lengths of the sides or hypotenuse of one of the resulting triangles.
- A 30-60-90 triangle is half of an equilateral triangle, so at times you'll need to slice an equilateral triangle in half, and then calculate from there.

Note: If you slice an equilateral triangle in half, be sure to remember to use the correct base in $A=\frac{1}{2}bh$. You will now have two possible bases—the base of the equilateral triangle, and the base of the 30-60-90 triangle (which is half as long). I see students confuse them all the time and lose points as a result.

EQUALLY IMPORTANT ARE the Pythagorean triples.

One of the dark secrets of the SAT is that you rarely need the famous Pythagorean Theorem, $a^2 + b^2 = c^2$, which is about the only thing anybody ever remembers from geometry class.

Focus instead on the Pythagorean triples, which save you a lot of time.

3 – 4 – 5
5 – 12 – 13

In other words, $3^2 + 4^2 = 5^2$ and $5^2 + 12^2 = 13^2$.

Memorize these triples. Tattoo them on your forearm.

Use them as your locker combination. They will speed up your performance on the math section.

Also, the bastards like to multiply the Pythagorean triples by a factor of 2, so memorize these as well:

6 – 8 – 10

10 – 24 – 26

And by a factor of 3:

9 – 12 – 15

15 – 36 – 39

Commit them all to memory, understand the special right triangles, and you should be good.

10

Rule #8: Memorize the Third-Side Theorem and the Three-Part Circle Formula of Awesomeness

It's easy to find the range of lengths of one side of the triangle, if you have the other two side lengths.

The length of any side of a triangle is always greater than the difference but less than the sum of the other two sides.

For example:

What is one possible length of the third side of a triangle with sides measuring 4 cm and 7 cm?

a) 2 cm
b) 3 cm

c) 9 cm
d) 11 cm

It's easy to work out. The difference of the other two sides is 3 cm, and the sum is 11 cm. The third side must be between those two values. In mathematical terms,

$3 < x < 11$. That's why C is the answer.

I'd like to point out the trap answers, B and D. The third-side theorem doesn't say greater than *or equal to*, or lesser than *or equal to*. So you can't use those values.

As of this writing, the new digital SAT hasn't tested this concept yet, but that doesn't mean it won't in the future.

ANOTHER LITTLE-KNOWN formula that will save you at least once, maybe twice, on exam day, is what I call the Three-Part Circle Formula of Awesomeness.

When the bastards draw a circle with a wedge drawn into it (say, 120 degrees), that's when you whip this out. Behold:

$$\frac{Degree}{360} = \frac{Arc}{Circumference} = \frac{Wedge}{Area}$$

It's indeed awesome. Each of the three segments of this formula represents a part divided by a whole: the degrees, the length around the outside, and the area. You can pick any two segments of this equation that you need to solve a problem and make a proportion. Be sure you have 3 of the 4 pieces of necessary information. Solve for the fourth.

Tattoo this formula along with the others on your forearm.

[There's a corollary too. For advanced test-takers, the SAT occasionally will require you to remember the rule that the arc measure formed by an angle with its vertex on the circle is double the measure of the angle.]

11

Rule #9: Advice for squares and rectangles

THE BASTARDS LOVE TO catch you on technicalities, and one very simple trick that they keep in their back pocket is this:

A square is a rectangle.

It's true. Technically, a rectangle is a class of shapes—all of which have four perpendicular angles—and a square is one of several categories of rectangle. I've seen this trip up more than a few test takers, even those who were high scorers.

Second, do you remember my earlier advice? *Read every flipping word in the question.* It applies here. For some unexplained reason, our minds don't like the word *perimeter* (which is the circumference of any non-circle). When we read that word, our brains often see the word *area* instead. No idea why it happens, but it does.

Let's be clear: *perimeter* $\neq$ *area*. They are not equal. They are different.

Similarly, for three-dimensional figures, our minds read the word *surface area* and change it to the word *volume*.

Let's be clear on this too: *surface area* $\neq$ *volume*.

Be on guard. The bastards are always looking for new and inventive ways to mess up your score.

12

Rule #10: Coordinate Geometry and Functions

THAT LOVELY LITTLE page of formulas that sits at the beginning of every SAT math section like a birthday present? It's missing one thing.

Coordinate geometry.

Funny thing is, there's no reason why the bastards don't include those coordinate geometry equations. They could, if they wanted to, but they are arbitrary and cruel, and that's the way life is.

What this means is that you'll have to remember a few more formulas. In order, here are the equation of a line, slope, and midpoint equations:

$$y = mx + b$$

$$\frac{y_2 - y_1}{x_2 - x_1}$$

$$\frac{x_1 + x_2}{2}, \frac{y_1 + y_2}{2}$$

You should already know how to use these, so I won't go into more detail or explanation. It's just more stuff to tattoo on that left forearm, which should be starting to look like a sleeve by now.

Also:

- parallel lines = equal slopes
- perpendicular lines = negative reciprocal slopes

There are **no** solutions if the two lines have the same slopes but different y-intercepts. There are **infinitely many** solutions if the two lines have the same slopes and the same y-intercept.

A lot of test takers get scared when they see functions, especially the mathematically averse. That's just fear. At the easy and medium questions, they're very doable. Just recall this very simple fact:

$f(x) = y$ is the same as (x, y)

In other words, if you see f(3)=7, you should instantly

write (3,7) on the paper. It's a point on the line. This is really the basis of all coordinate geometry on the SAT.

Typically, the bastards will blend functions with coordinate geometry:

If $f(x)$ is a linear function such that $f(0)=4$ and $f(2)=10$, what is the slope of the graph of $y=f(x)$?

a) 0
b) ½
c) ¾
d) 3

They gave you two points on the line. Did you see them?

F(0)= 4 means (0,4).

F(2) = 10 means (2,10).

Now use the slope formula. 10 - 4 = 6 and 2 - 0 = 2. 6/2 = 3.

Slope is 3. The correct answer is D.

There are many more complicated variations on this, but those questions are primarily directed at students aiming for 700 or greater. If you're aiming for a 600, this is enough.

13

Rule #11: Memorize the growth and decay formula.

It's not given to you, but this formula is necessary. You need to merely recognize it in some questions, and use to solve for the answer in others.

The growth and decay formula is mostly used to describe the growth of money and bacteria. (Ironic that they're both measured in the same way, isn't it?) These are usually word problems involving financial investments or laboratory experiments using petri dishes.

Here's the formula—

$$\text{Total} = (\text{initial amount})(1 +/- \text{rate of growth/decay})^{\text{number of time intervals}}$$

For example:

A truck is purchased for $50,000. Its value depreciates by

10% annually. After how many years will it be valued at $40,500?

a) 1
b) 2
c) 3
d) 4

Put in the values that you know. $40{,}500 = (50{,}000)(0.90)^{X}$.

Now plug in the answers.

When x=2, the equation in fact yields 40,500. The correct answer is B.

14

Graphing Functions

REMEMBER the following rules about graphing exponential functions.

- Any function with x^1 is a straight line (0 turns).
- Any function with x^2 is a U-shaped function (1 turn). Also called a parabola.
- Any function with x^3 is an S-shaped function (2 turns).
- Any function with x^4 is a W-shaped function (3 turns).

On graphing function questions, remembering these rules alone will often help you eliminate 2 of 4 answers.

Furthermore, for any question that asks something like *for how many values of x does f(x)=9*, just draw a straight line at the given value. In this example, since f(x)=y, you know that y=9. Draw a horizontal line that crosses the y-axis at 9. Then count the number of times that line crosses the function. That's the number of values of x.

15

I'm all about giving value, so here is a list of the concepts and equations you need to know for the digital SAT math section.

This list includes both basic and advanced mathematical concepts that you need. Some are used in both types of questions.

Percent Change:

$$Percent\ Change = \frac{Difference}{Original} x\ 100$$

The quadratic equation:

$$ax^2 + bx + c = 0$$

. . .

THE THREE QUADRATIC IDENTITIES:

$$(\mathbf{x}^2 - \mathbf{y}^2) = (\mathbf{x} + \mathbf{y})(\mathbf{x} - \mathbf{y})$$
$$(\mathbf{x}^2 + 2\mathbf{xy} + \mathbf{y}^2) = (\mathbf{x} + \mathbf{y})^2$$
$$(\mathbf{x}^2 - 2\mathbf{xy} + \mathbf{y}^2) = (\mathbf{x} - \mathbf{y})^2$$

THE QUADRATIC FORMULA:

$$x = \frac{-b \pm \sqrt{b^2 - 4ac}}{2a}$$

Know what the discriminant (the part inside the root) is.

- If the discriminant is **positive**, then there are **2 real roots** (solutions).
- If the discriminant is **zero**, then there is **1 real root** (solutions).
- If the discriminant is **negative**, then there are **0 real roots** (solutions).

DIRECT PROPORTION:

$$\frac{a_1}{b_1} = \frac{a_2}{b_2}$$

INDIRECT (INVERSE) PROPORTION:

$$a_1 b_1 = a_2 b_2$$

PROBABILITY:

$$Probability = \frac{Wanted}{Total}$$

AVERAGE:

$$Average = \frac{TotalValue}{Numberofthings}$$

Also review weighted averages. This occurs when you have a population divided into two different groups.

. . .

Distance:

$$Distance = Rate\ x\ Time$$

Surface Area of a Cube (where *s* represents a side length):

$$Surface\ Area\ of\ a\ Cube = 6s^3$$

Systems of Equations/Substitution:

Surely you've seen something like this before—

$$4a - 6b = 18$$
$$3a + 3b = 30$$

To solve, you have two choices.

- **Substitution**: You can solve for one of the variables, then pop that back into one of the equations and solve for the other.
- **Systems**: First, you can stack them on top of one another. Then multiply one of the equations in such a way that one of the variables' coefficients is the same as the coefficient of that same variable in the other

equation. Then add/subtract both equations together to cancel out that variable.

Completing the Square:

This is too complicated to describe here. It's also an advanced technique, best memorized by those trying to get a perfect score. If you want 750 or higher, be sure to review it.

The Vertex (*h,k*) Form of a Parabola:

$$a(x-h)^2+k$$

Recall also that the vertex of a parabola is located at the midpoint of its x-intercepts.

Equation of a circle with center (h,k) and radius r:

$$(x-h)^2+(y-k)^2=r^2$$

Similar Triangles:

The measurements of the 3 angles of similar triangles are the same for both triangles. In other words, if the degree measurements of the angles of one similar triangle measure 50-60-70 degrees, the degree measurements of the angles of other similar triangles will also measure 50-60-70 degrees.

However, while the 3 sides of similar triangles don't have the same measurements, they do all have the same proportional lengths. In other words, if the three sides of one similar triangle measure 5 cm, 12 cm, and 13 cm, then the three sides of other similar triangles could measure 10 cm, 24 cm, and 26 cm (factor of 2). Or they could measure 15 cm, 36 cm, and 39 cm (factor of 3).

Intersection of Two Lines:

Repeat after me: To find the intersection of two lines, set the equations equal to one another. Solve for x. Then use substitution to solve for y.

Absolute Value Inequalities:

It's a basic trap, but remember to reverse the inequality sign if you multiply or divide by a negative number. The bastards have been putting this on the exam for decades.

16

A word about trust.

You may be hyperalert now to the fact that the bastards are trying to screw you. They definitely are. The good news, though, is that the bastards usually stop themselves from flatout *lying*.

Usually. There is one exception.

Their figures.

When the bastards provide a sketch of a geometric figure, *don't trust its proportions*. And don't try to somehow measure the drawing either—you know, the desperate thing where you count three and a half little finger segments, or two pen caps.

Why? Because the figures are wrong.

In other words, the sides of the figure are not proportional to each other. It's a little like that sentence that used to appear on rearview mirrors: *Objects in mirror are closer than they appear*. In fact, sometimes the bastards even acknowledge their bastardlyness. Have you noticed that occasional *Note: Figure not drawn to scale* underneath a shape? Assume that is true for all the figures.

Parallel lines are another favorite trick. Remember, *parallel lines are only parallel if the exam tells you they're parallel.*

The lines may *appear* parallel in the figure, but that means nothing. It has to be explicitly stated through—

1. The word *parallel*
2. The parallel symbol (double vertical sticks)
3. Being part of a figure that by definition contains parallel lines (square, rectangle, parallelogram, etc).

However, you don't have to be any more paranoid beyond these things I've just described. For example, you can trust is that a straight line is a straight line, that a circle is a circle, that a perpendicular angle is a perpendicular angle (if the little marker is present), etc.

17

There is obviously much more to say about the SAT math modules, but the purpose of this handbook was to provide low- and medium-scoring test takers with a concise, expert way of raising their scores to 600.

So how should you study?

There is no one right way to study for the SAT. Some students like to take a diagnostic practice test immediately. Others want to learn the material before subjecting themselves to an exam under test-day conditions.

My suggestions:

1. **Start slowly**. Don't worry about your timing right away. It's like studying a new piece of music for an instrument: learn the fingerings first, and the speed comes later.
2. **Work on a few questions at a time**. I suggest 25- or 30-minute segments. This is the Pomodoro method. Take baby steps, and soon you'll be walking.

3. **Check the explanation to each question right away**. Strike when the iron is hot! Learn when your thoughts are still fresh.
4. **It's good to get questions wrong**. This tells you exactly what you need to improve.
5. **Keep a log**. Every time you miss a question, write down a description of its type, e.g. graphing functions. As the mistakes add up over time, you'll see the patterns. This is gold.
6. **Don't take too many full practice tests**. I'd limit to 4, maximum. In fact, don't always practice with time pressure, at least not until shortly before the exam.
7. **Use real SAT questions whenever possible, via the Bluebook app**. If you run out of practice questions, use third-party materials.
8. <u>Recommended</u>: Khan Academy, Princeton Review, and Ivy Global. All of them are excellent.
9. <u>Not recommended</u>: Kaplan, McGraw-Hill, Barron's, Peterson's. Their materials are substandard or inaccurate and will hurt your performance.
10. **After a practice exam, take the rest of the day off.** Don't bother trying to review your test results immediately afterward. Your brain will most likely be fried. Wait until the next day, when your mind is fresh, to review your results.
11. **Don't give up.** Most people have to take the SAT two or three times. It's not a problem.

These days, many colleges will form a
superscore for you anyways.

MOST IMPORTANTLY, the act of taking the practice test itself is only a prelude. What leads to actual improvement is the careful review of each question you got wrong.

And remember this when you're feeling down: the SAT Math modules don't measure anything except your ability to take the SAT Math. They don't predict future success, not in anything.

But it's still good to try your very best at everything in your life, because excellence is a habit.

Now let's move to the Reading and Writing modules.

READING

18

There are two Reading and Writing modules on the digital SAT. In each, you are given 32 minutes to complete 27 questions. So that's a total of 54 questions in 64 minutes.

In other words, you have 71 seconds per question. Please write that down somewhere.

The first module contains a mix of easy, medium, and hard questions. However, the adaptive nature of the new digital SAT means that your second module will either be more difficult or get easier, depending on how you performed on the first module.

Nervous yet?

ETS has been using this model for a decade already with a different exam, the GRE, which is used by a lot of graduate school programs. Now the bastards have basically copied the same format to the high school exam. I'm surprised it took them so long to do it. I expected a digital SAT back in 2015, but it never materialized.

On the Reading and Writing exam, there is only one question format: multiple choice. There are four answer choices for each question. Everybody knows multiple choice, so the less said, the better.

Lastly, two questions in each module are experimental —that's four questions total—and don't count towards your score. You won't know which ones are experimental, and don't try to guess which ones they are either. Treat them all the same.

HERE's something else about scoring.

It's possible that you're just not very verbally inclined. Maybe you've never liked looking at words on a page. Maybe you feel more comfortable with numbers. Maybe you eat tables of data for breakfast. Maybe Tuesday-afternoon chemistry lab is your jam because you love to work with your hands.

That's fine. I've literally helped over a thousand students like you, some of whom were totally paralyzed by low self-esteem as soon as they so much as thought about a paragraph. Don't worry about that. I would urge you to remember, when you feel like you're not good enough, that we all can't be good at everything—and colleges know this better than anybody. They don't expect perfection from you, no matter how much your parents and counselors may have tried to brainwash you.

In fact, if you're planning to go into a STEM field, here's a radical piece of advice: Just show colleges that you're *decent* with words.

You don't need to show them that you're brilliant. Just give them at least one reason to accept you—and don't give them any reasons to reject you.

The most common type of student I've worked with is a

child of an immigrant family who is planning to enter a STEM field. I've lost count of the number of students with this profile I've helped. These immigrant families push their kids hard, especially the East Asian ones, but the one thing those parents overlook is the fact that their student is fighting an uphill battle with the English language. Across cultures, a person can learn math much more easily than a person can learn reading and writing.

Put it this way: Sometimes the student has trouble with English because he or she was born in a foreign culture. Or, more commonly, the student was born in the United States but has been raised to communicate with his or her parents in their native language while speaking English outside of the house. That's fine, but in either situation it's silly to expect that student to achieve perfection in English class. We learn languages first and foremost at home, and if your parents don't speak the English well, or at all, it'll affect your own English greatly. It's also not fair to those students to expect them to get a top Reading and Writing score on the SAT—though a few manage to do it.

Me, I can quickly teach people English grammar. It's easy, it's a straightforward system, and with a minimal amount of practice you can get every question correct. Rigid thinkers who are going into engineering, for example, respond well to the rules of grammar. For them, reading is a different matter, for a few different reasons. However, with such short texts—only one paragraph each—it's also become easier to improve one's reading score on the SAT than ever before.

Anyways, if you're an immigrant student planning to become a STEM major, don't worry about getting a perfect score on Reading and Writing, or even scoring a 700. It's not necessary. College admissions committees don't punish you for what you're lacking—they celebrate you for what you *are*. Show excellence in other stuff, do well enough on Reading and Writing, and more often than not, they'll show mercy.

With that said, a good goal for everybody who struggles with Reading and Writing is this:

A 600.

This portion of this handbook has been written with that minimum goal in mind. Higher is better, but make sure you can get at least a 600. After that, aim for a 700.

And the best part of this deal? Scoring a 600 on the Reading and Writing isn't that hard.

Someone once said that it's wise to learn from your mistakes, but it's even wiser to learn from the mistakes of others. In my career as a tutor of standardized testing, I've watched people commit the same mistakes, stumble into the same tiger pits, over and over. Through long experience, I have become able to predict the mistakes you are going to make long before you make them.

This portion of this handbook is a list of those errors, explained in patient detail. I want you to learn from the mistakes of others.

It begins in the next chapter.

19

First off, the world wasn't kidding when it told you to read.

I'm not talking about reading today, when you're in the tenth or eleventh or twelfth grade, when you're loaded with more textbooks than your poor Jansport can carry. In fact, if you've challenged yourself with hard classes and sports and extracurriculars, you've probably seen your *happyfunreadingtime* (my made-up term for pleasure reading) decrease throughout high school.

No, I'm talking about when you were in elementary school and junior high. You should've been reading everything—magazines, newspapers, backs of cereal boxes, recipes, instruction manuals, skywriting, discarded Post-It notes, tattoos, tea leaves, and any other form of text that passed across your eyes.

[If you didn't, here's what to do: Help someone else avoid the same mistake. Go to your little brother or sister or cousin or neighbor, find some random third grader, and demand that they read at least one young adult book every week. Find a way to enforce that order. You can't change your own past self, but you *can* pay it forward.]

Now, you're probably also assuming that I'm talking about reading fiction. Not at all. Even though *Harry Potter and the Grotto of Turnip Fairies* may be your all-time oh-my-God favorite book, YA (young adult) and fantasy novels don't directly help your comprehension of nonfiction material. Most of the reading paragraphs on the SAT are nonfiction, and they're drawn from college-level material, which means twelfth grade and beyond.

The best material to read in preparation for the SAT are sophisticated magazine articles. I always recommend magazines such as *The Atlantic*, which is America's oldest and possibly its best magazine (though it does skew to a slightly older crowd, so maybe you want to hold off on that). NBA and NFL and MLB fanatics can subscribe to *Sports Illustrated*, whose articles are illuminating and sophisticated. Future scientists can read *Scientific American.* In fact, until the nineteen-eighties, the bastards actually plucked articles straight out of the pages of that magazine and reprinted them in the SAT.

Furthermore, all those years of scanning this type of text will have literally changed the way you read. Experts understand that we absorb writing in chunks of various sizes from small to large—letter by letter, word by word, phrase by phrase, clause by clause, sentence by sentence, paragraph by paragraph. The higher you are on this great chain of reading, the bigger the chunk, and the better a reader you are. Speaking personally, I've noticed that my own chunks are now the size of half a paragraph, but it took until my mid-twenties to achieve that.

Another reason that you want to increase the size of your "chunking" is that in college, you will be given more reading assignments than you can possibly accomplish, especially if you're still doing the finger-under-the-words-saying-them-on-your-lips thing. For example, you can't possibly read eighty pages of poorly translated polemics by fourth-century church historian Eusebius in one evening when you have three other assignments to finish. (Yes, that's my personal experience talking right there.) Instead, you have to chunk. In fact, most liberal arts college students are forced to scan and annotate, and those are other skills you can use on the critical reading portion of the test.

And your Reading and Writing score actually does predict your freshman year GPA in college with some accuracy.

Let me guess what you're thinking: *It's not that I hate reading—it's that the passages are so booooring.*

I can understand that frustration, and I have a simple solution.

Pretend like you're hanging out with your parents' friends.

We know what that's like. Remember sitting disconsolately at some restaurant, swinging your feet under the table, hearing the adult voices get louder and more off-color as the wine in the glasses got lower and lower? Remember listening to the voices yammer about stuff that didn't interest us—interest rates, doctor's office visits, FICO scores, what's happening with the gang from the old neighborhood? Remember laughing politely at the jokes that weren't very funny?

The reading paragraphs are like your parents' friends. However, while those dinners lasted for hours, you only need to be pretend to be interested in these paragraphs for about a minute each.

Here's another perspective: Have you ever been forced to pretend to like someone at school? Most of us have. It's common knowledge that girls in particular often have, shall we say, *fluid* friendships. (*Frenemies* is the better word.) It's sad, but being fake is required of most of us at least once in a while, and it's not necessarily harmful.

Approach the reading like that.

20

The reading portion of the exam is difficult because it's a gray area. Not heather gray like a college sweatshirt, but gray as in *murky*. The reasoning for the answers is unclear, somewhere between black and white.

Math, on the other hand, has definite answers—no mystery there. Grammar too.

Not so for reading.

Some people can't handle this. These are people who hate uncertainty. They organize their socks in neat lines. They straighten the items on their desks at perpendicular angles. They used daily planners at six years old.

So when those types of people are confronted with the reading section, they go blank. They can't cope. They need a system.

I will give you that system near the end of this handbook, but first let's look at how, without expert guidance, those insecure test-takers will incorrectly prepare for the SAT.

They study vocabulary.

It's their security blanket. Memorization makes them

feel like they're accomplishing something tangible. They memorize page after page, sheaf after sheaf, of vocabulary words such as *abrogate* and *contusion* and *lachrymose*.

That's all well and good, and I admire the effort, but it's also misguided. For many reasons.

One, they probably won't remember more than ten percent of the words. Be honest: how long do you remember vocabulary after you crammed it into your short-term memory for a quiz in English class? Expert psychologists call this the active, or primary, memory, and they've noted that we can't really keep any more than seven items in that part of our brain at one time, and that these items don't stick around for more than about twenty seconds.

[Side note: The Magical Number Seven is a real thing, and it came from the title of a famous paper by psychologist George Miller. The main idea: We humans get confused when presented with more than seven items to choose from. Just watch people at a restaurant when confronted with a huge menu. Often, unable to decide what to order, they ask other diners for help, or ask the wait staff what's best. That's why the most attractive, manageable menus usually have no more than seven items. I also wonder if that's how we humans settled upon a seven-day week.]

Two, the bastards—you remember the bastards, right? the writers of this test?—already know that students memorize only the first definition of a word. So, summoning all their bastardliness, they often choose *connotations* (second meanings), not *denotations* (first meanings).

Here's an example: the word *flag*. What does it mean? You probably thought of the Stars and Stripes, flapping proudly. That's the denotation. However, one of its connotations is "to lose energy." For example: *My energy flagged near the end of the triathalon.*

I didn't make that up. It was on an SAT years ago. The bastards play dirty.

Most crucially, there is no way to study all the words the SAT can possibly throw at you. Linguists estimate the number of words used in common English conversation to be about 175,000—and our language contains about 750,000 total, if you include technical and obsolete words. Given those odds, chances are that you're not going to even make a dent in the English language. Those of you who persist in attempting to memorize thousands of vocabulary words in the face of such odds could probably use some couch time with a very friendly therapist.

However, there is a list of words that matters—the ones that have been carefully curated. That means finding someone who went through lots of publicly released SATs over the past several years and carefully compiled a list of the most commonly used words.

I've included such a list at the back of this handbook. You're welcome.

21

I don't mean to sound too negative about vocabulary. It's not a lost cause, and a bigger vocabulary leads to all sorts of wonderful things later in life, such as fame, happiness, and adoration from attractive members of whatever sex you find attractive.

Joking, joking. But a good vocabulary *can* lead to better jobs and more money, when you learn how to use the right word at the right moment. That's called *congruence*, and when you deploy that perfect adjective at that perfect gap in conversation, your listener will respect you, and even like you more. This is why you want to have those words in your back pocket.

And yes, there are short-term things you can do to boost your vocabulary.

First, vocabulary is not learned in a vacuum. We remember it better when it touches other stuff.

Here: Which one is easier for you to remember?

Tormentor (noun): A person who afflicts another with great bodily pain and suffering.

Or this—

After my sister's tormentor shoved her into the trees, she used her brass knuckles to turn his face into bloody oatmeal. Fist bump, sis.

Most people agree that making example sentences is more effective than memorizing plain white-bread definitions.

Here are some methods of inserting context into vocabulary.

1. **Sample sentences.** If you're going to attempt memorization, okay, but for God's sake don't forget everything I just said and start writing definitions. Instead, write a sample sentence—and the more vivid the better. (See brass knuckles and face-oatmeal imagery above.) Be sure that the context tells you the meaning of the word too; it's no use writing sentences like *The man was arrogant*. Instead, write *The arrogant man peered down at me as though I were a pimple on the ass of humanity*. It's more fun that way, and you get bonus points for turning yourself into a more descriptive writer too. However, this is time-consuming.
2. **Vocabulary novels.** Yep, they exist, and you're probably kicking yourself for not having read them before. Some enterprising publishing companies have commissioned mysteries and adventures to be written with ten bolded vocabulary words on every page. There are

usually review quizzes at the end of each chapter. I've read a couple and can highly recommend them as fun ways to expose yourself to more vocabulary. One good title *is Simon's Saga for the SAT* by Phillip Geer. Also, check out Kaplan's SAT Score-Raising Classics, in which they put a classic novel on the left pages and help you with weird or difficult words on the right pages. They've done *War of the Worlds*, *The Scarlet Letter*, *Wuthering Heights*, and others this way. It's a great way to read older classics that might otherwise scare you off.

3. **Vocabulary pictures.** Learners can be divided into those who are visual (seeing), auditory (hearing), and kinesthetic (doing). For those of us in the first category, get your greedy little hands on a book of vocabulary cartoons. They connect tough vocabulary words with pictures. For example, *fickle* will feature "fickle pickle" as a rhyme, followed by a funny cartoon of a pickle trying to make up its mind in a grocery store. It works well. There are several books like this, but *Vocabulary Cartoons: SAT Word Power* is a good place to start.
4. **Vocabulary apps.** If, like the rest of the western world, your life revolves around your phone—and we totally need a new term for this type of device, since nobody actually talks on them anymore—then maybe you want to play picture games with words. SAT Vocab by Mindsnacks is one such app, and Brainscape and Anki are also good choices.

5. **Vocabulary email.** If you really love receiving emails, then sign up for the many and various Word of the Day services available. They're free, and they're better than sitting around cracking rocks. (That's something my grandmother used to say. Let me if you know what it means, because I don't.)
6. **Latin and Greek word roots.** Years ago, in a school system far, far away, all high school students were required to take two years of Latin. That wasn't so they could speak Latin, which hasn't been spoken for centuries. It was to improve their English vocabulary. See, two thousand years ago, Latin and Greek gave birth to five children: Spanish, Portuguese, French, Italian, and Romanian. They also birthed an illegitimate stepchild known as English, which was mocked so mercilessly for its lowly half-Germanic origins that it developed very low self-esteem, and then wildly overcompensated by becoming the business language of the entire modern world. Anyways, if your English teacher has a clue, he or she has already shown you some of these Latin and Greek word roots. For example, you will have already learned that *synchronize* is actually composed of three roots: *syn-* (meaning "same"), *chron* (meaning "time"), and *–ize* (meaning "to do or make"). Putting them all together shows you that *synchronize* means "to do at the same time." If you never learned these roots, you should sue your school district. I've included a list of Greek and Latin

word roots in the back of this handbook. Once again, you're welcome.

Now let's take a look at the 10 types of questions that you will encounter on test day.

22

The 10 Types of Questions on the Reading and Writing Section

These are listed in order of their appearance.

Craft & Structure – Reading – First 6 to 8 questions

- **Question Type #1: Words In Context**
- **Question Type #2: Text Structure and Purpose**
- **Question Type #3: Cross-Text Connections**

Information & Ideas – Reading – Next 6 to 8 questions

- **Question Type #4: Central Ideas & Details**
- **Question Type #5: Command of Evidence**
- **Question Type #6: Inferences**

Standard English Conventions – Writing – Next 5 to 8 Questions

- **Question Type #7: Sentence Boundaries**
- **Question Type #8: Form, Structure, and Sense**

Expression of Ideas – Writing – Last 4 to 6 Questions

- **Question Type #9: Transitions**
- **Question Type #10: Rhetorical Synthesis**

The Reading and Writing modules follow a predictable structure. You'll see questions grouped by type and then ordered by difficulty. For example, if you have 3 words-in-context questions, they'll appear in order of difficulty, from easiest to hardest. Then, if you have 3 cross-test connections questions, they'll appear in order of difficulty too.

Keep this in mind as you figure out your strategy for the Reading and Writing modules. Predicting both the order of question types and the difficulty of question types can help you can better use your time on the exam!

23

Question Type #1: Words in Context

As mentioned earlier, the ancient Chinese general Sun Tzu told us to "Know thy enemy". That's good advice. Most of my students have little to no awareness of where they are inside a test, how much they've done, and how much is coming. It's time to fix that.

The Words in Context questions should always arrive first in the Reading and Writing module. Therefore, you need a technique to address them. Here's what a typical one looks like:

Unlike people in advanced societies with market economies, people in less developed societies still depend on subsistence agriculture for survival. These farming communities grow their own food, with neighbors bartering for other goods they can't produce themselves. In the past, these people also typically wore _____ clothing, since they were making their own garments as well.

a) soiled
b) old
c) organic
d) rustic

YES, Adimu is a real name. The College Board doesn't use hyphenated-American groups (African-American, Vietnamese-American, etc) because of the political controversy surrounding them. But the bastards still want to appear diverse, so they inject different ethnicities into the exam through names. This also helps them avoid discrimination lawsuits.

Anyways, the worst thing to do is probably the first thing newbie test-takers instinctively do: plug each answer choice into the blank, reading each on their lips to see how it sounds in the sentence. That's not only a huge waste of time, especially for slower readers, but it also leaves you exposed to a wide variety of verbal tricks designed to shaft you. For example, the bastards will often use homophones (*meddlesome* v. *mettlesome*) to throw you off the scent.

Here's a better way to approach:

1. Cover the answers.
2. Read the sentence or paragraph.
3. Look for the keywords.
4. Look for the transition words, which either keep the sentence going in the same direction (*and*, *furthermore*, *moreover*, etc) or turn it in the opposite direction (*but*, *yet*, *however*, etc).

5. Use steps 3 and 4 write down a *prediction word.* A good one here would be *homemade.*
6. Uncover answers. Eliminate any answer that doesn't match with your prediction. Then choose.

Note: The prediction, step #5, is the most important. Think of your own word to go in the blank. It can be a tiny one, like *mud* or *butt*, or it could be a big one, like *anachronism*. You could even recycle the key words in the sentence —I recommend this whenever possible, as it requires less thinking. You could even think of your favorite dancing turd emoticon, if that has a special meaning for you. Whatever you choose, just make sure the *meaning* is spot on. The part of speech doesn't matter; neither does syntax.

The correct answer is D.

If you're like most students, you've heard lots of urban myths about guessing on the SAT. You've probably heard that you should only guess if you can get down to two answers.

That's not even remotely true.

The truth is this: Answer every question on the exam.

24

Question Type #2: Text Structure and Purpose

THESE QUESTIONS ASK you to analyze how a text is put together, its overall purpose, and the purpose of its individual components.

Here's an example:

THE FOLLOWING text is an excerpt from The Oxford Diaries *by J.A. Jernay.*

Some people believe that the famous playwright William Shakespeare never came from Stratford-upon-Avon, that it's all a farce. There is some evidence to back up this point of view: documents from the time show that William Shakespeare in Stratford was a farmer who never spelled his name the same way twice. I like a different theory—that fellow playwright Christopher Marlowe faked his own murder to escape the King's agents (who wanted to kill him for having an affair with the king's mistress) and continued to publish his plays from abroad under the name William Shakespeare, an illiterate actor from the sticks. Most probable, though, is the possibility that Shakespeare wasn't one person at all, but a group of writers.

Which choice best states the main purpose of the text?

a) To question whether Shakespeare was truthful to those around him
b) To criticize the gullibility of mass audiences
c) To present different theories of Shakespeare's true identity
d) To reconcile opposing ideas regarding Shakespeare's authorship

The correct answer is choice C.

As mentioned above, these questions can ask for the main purpose of the paragraph, or for the purpose of a single sentence. Either way, be sure to select the safe answer.

25

Question Type #3: Cross-Text Connections

These questions ask you to analyze the relationship between two very short texts. Together they will total less than 150 words, same as a single text.

Here's an example:

Text 1

What type of factors have influenced the rise of the digital nomad community? The biggest one is the enhanced wifi networks now found around the world, coupled with powerful and portable laptop computers. Another is the willingness of more employers to allow contract workers to live where they would like. Others include cheap flights and ease of renting housing in foreign countries through websites such as Airbnb.

. . .

Text 2

Digital nomadism has been growing in popularity for a decade or more, and it experienced a huge surge in popularity since the covid-19 pandemic. However, lately more employers are starting to question the wisdom of an all-remote workforce. Several have begun pulling back from the movement and are now requiring employees to return to the office at least three days each week.

Based on these texts, how would the author of Text 2 most likely describe the view of the author of Text 1?

a) It is correct in the big picture but incorrect in the minor details.
b) It accurately describes a past scenario, but it hasn't taken into account recent changes.
c) It holds true only under certain conditions.
d) It doesn't acknowledge the variety of individual experience.

How to do these?

First, read each one and boil each down to its stupidest possible main idea. Your summaries should be so simple that they sound a bit stupid.

Imagine how you would explain the passages to a five-year-old. That's what you should aim for.

Second, determine the relationship. Typically there are three types of relationships:

- same (supports)
- opposite (criticizes)

- perpendicular (maybe partly agree, maybe no relationship at all)

Third, select the choice that has the same relationship as the one you decided upon.

The correct answer is B.

26

Question Type #4: Central Ideas & Details

THESE QUESTIONS FOCUS on identifying main ideas and supporting details.

Here's an example—

Robert Canavan was the head of a prestigious geography department at a university on the East Coast. It looked as though he'd spent the last thirty years slumped behind a desk. His body was shaped like a question mark. He wore a black blazer, a light blue Oxford button down shirt, shapeless beige slacks, and brown tassel loafers. Cosmo had met him at this same conference the previous year, and Robert had instantly distrusted him. Cosmo assumed it was because his own fitness level was a physical reminder to Robert of what he'd lost over the years.

Which choice best states the main idea of the text?

a) Cosmo thinks Robert is jealous of him because of his physical health.
b) Cosmo assumes Robert has had medical issues in the past.
c) Cosmo usually finds reasons to create conflict with colleagues.
d) Cosmo responds to most human relationships in a cynical manner.

Much like the Cross-Text Connections questions, you should boil down this text to its simplest possible main idea. Again, if you can explain it to a five-year-old, you're on the right track.

Then find the answer choice that presents that same idea in a positive, non-judgmental manner.

The correct answer is A.

27

Question Type #5: Command of Evidence

There are two types of this question:

- text evidence
- quantitative evidence

Text evidence questions ask you to identify a finding from a study or a quote from a given work that supports a given claim.

Quantitative evidence questions ask you to find evidence in a graph or table to support a given claim.

Here's an example—

Sam Dewey, Jo Cheatem, and Gordon Howe have hypothesized that the reduction in associate positions at legal firms has been driven primarily by technological changes, not by changing beliefs about the legal profession, or any other factor. To test their hypothesis, they sent out a survey to hiring managers at seven hundred legal firms, asking them to select at least two reasons for the recent reduction in hiring. The survey supplied seven possible reasons to select from.

Which finding from the surveys, if true, would most strongly support Dewey, Cheatem, and Howe's hypothesis?

a) At least 20% of the respondents selected technological changes as a reason.
b) At least 50% of the respondents selected technological changes as a reason.
c) The respondents selected technological changes more often than any other reason.
d) The majority of respondents declined to select more than one reason.

You don't need any outside information for this type of question. Identify the argument, stay specific, and create a simplified version of the hypothesis for yourself. Read each choice while keeping this dumbed-down hypothesis squarely in the front of your mind.

Remember that advice about "make predictions wherever possible"? I really meant it. You should be predicting every answer to every Reading question before it's made.

The correct answer was C.

(The trap answers were A and B. Because respondents were asked to select at least two reasons, neither of those choices eliminates the possibility that the respondents selected another reason more often than technological changes.)

28

Question Type #6: Inferences

Inference questions ask you draw a conclusion based on information in the text. They all use the verbs *infer, imply, conclude,* or *suggest.*

For example—

It can be inferred from the discussion of land sharks that...
In lines 4-5, the author implies that...
The author suggests in line 7 that...

Your task is simple. You, the annoyed test taker, need to guess what the author would conclude, if only he or she were here to tell you.

This is always difficult. It can be easier, though, if you remember a simple piece of advice:

Choose the safe, obvious, "dumb" answer.

Let me give you a hypothetical dialogue that illustrates

just how sensitive and careful you have to be in choosing an answer.

My cat died today. What can we infer?

Q: *Can we infer that he was old?*

No. It could've been struck by a car.

Q: *Can we infer that he was sick?*

No. It could've been sick with liver disease.

Q: *Can we infer that he was killed?*

No. It could've died of old age.

Q: *Can we infer that I'm sad?*

No. Be careful here. Just because *you* love *your* cat doesn't mean that *everyone* does. At the end of his life, my grandfather hated his cat and swore to outlive it. Unfortunately, the cat won that contest.

Q: So what can we infer?

We can infer that the cat was alive before it died.

How stupid does that sound? *The cat was alive before it died.* Duh! However, it's the smallest possible extrapolation we can make from the four words *My cat died today*. It's an unrefutable statement. And if you like this playing this type of mental game, you should definitely go to law school.

Once again: *Choose the safe, obvious, "dumb" answer.*

29

The Six Types of Wrong Answers on Reading Questions

For many students, the Reading questions in the preceding chapters are the most difficult ones on the entire exam.

Why? There is no clear-cut answer, no one hundred percent certainty, no *ah-ha* moment during which a pair of fists gets thrown into the air while the opening chords of "Don't Stop Believin" fill the arena.

That means one thing: *If you can't always pinpoint the right answer, then at least you should eliminate wrong ones.*

It's a subtle distinction, but eliminating wrong answers is the best strategy for this part of the test. Weed out, weed out, weed out—and whatever answer remains gets chosen and lifted high into the air, off to a better place.

The good news: there is a system of rules regarding bad answers. You can use this system of rules to weed out the undesirables, which will help you get closer to the right answer.

This system of rules is my own. I invented them over the last couple of decades of trying to explain the SAT to a couple thousand students. These rules have been refined to as sharp a point—and in as entertaining a manner—as can possibly be done. *And they work.* Many students have thanked me for them.

I'm giving them to you now.

Remember, though, that these are merely guidelines. They work about 80 to 90 percent of the time. There will always be questions that are exceptions to the rule.

Here's the list—

1. *Too extreme.* Any answer that can be classified as a 0% or 100% word should be eliminated. The reason: Such statements are too easy to knock down. Watch out for the words *always, never, only, all, none, prove, comprehensive,* and *total.* You can eliminate these answers most of the time.
2. *Too negative.* Like the scheming cheerleader who smiles to your face but spreads wicked gossip behind your back, the bastards who write this test like to outwardly pretend that life is one long ride on the back of a unicorn through a sunlit field of lollipops. Accordingly, any negative answer will usually be wrong. Watch out for words such as *no, not,* and *never*. Eliminate negative emotions too, especially *frustrated* and *annoyed*. If the bastards ever do want to say

something negative, they'll often use the words *limited* or *misguided*, which are correct because they're softer. (One important exception: a negative word used with an extreme word, such as *not always*, usually indicates a correct answer, because something extreme is being reined in.)

3. *Too judgmental.* In philosophy, there is a long-standing debate between *absolute* and *relative* thought. Absolutists believe that there is an unchanging rule of standards by which everything can be judged. Relativists believe in a sliding scale on which nothing can be judged absolutely. But the bastards hate absolute, or judgmental, answers. See, if those answers were to be correct, the bastards would expose themselves to a hundred lawsuits from people who object to such judgments. So avoid any answer containing the words *better, best, worse, worst, superior, inferior,* or *should.*
4. *Predicts the future.* Nobody can see into the future (except for Warren Buffett), so trying to predict it is a dangerous business to your bank account and to your SAT score. Eliminate any answer containing the word *will.* The only exceptions are texts that are literally about some aspect of the future.
5. *Off topic.* If you've been awake at any point in your educational career, you've probably learned how to identify an off-topic answer, so there isn't much to say here.
6. *Misinterpreted.* These are the trickiest answers to identify. Here are three varieties.

a) Starts good, but ends bad. For example: *Climate change that caused polar bears to swim further to feed on unsuspecting beachgoers in Argentina.* Everything sounds plausible until the last four words, which are ludicrous. I wrote them to make you laugh, but the bastards won't be so nice—they make the end of a misinterpreted answer sound just as plausible as the beginning. Remember: *half good = all bad.* This is especially dangerous for students who don't like to finish reading the answer choices.

b) Reversal of cause and effect. Quickly read this sentence: *In 1492, Christopher Columbus sailed from America to Europe in search of a passage to India.* If you read it quickly, or if your eyes are tired, or if you don't give two drips of runny crap, this sentence looks perfectly okay. A second glance will show you, however, that the words *America* and *Europe* were switched. *A causes B* is translated to *B causes A*. Careful of these.

c) On general questions, the wrong answer will sometimes pluck a single detail from the passage and attempt to pass it off as the main idea.

IMPORTANT NOTE: The above six tips apply to the *answers only*. It's possible, but not likely, that the bastards would include a text that contains judgmental comments, negative dialogue, and extreme words. Regardless, the correct answer choice will avoid such language.

Memorize this list. Recite it to yourself. Apply it in your SAT practice. I quiz my students on it, which provokes some rolled eyeballs, but their scores nearly always rise. You should see a noticeable uptick in your score. It might only be fifty points, but every little bit counts.

30

The Six Types of Right Answers on Reading Questions

By contrast, the types of *right* answers on passage-based reading are more fluid. They also don't always conform to a consistent look or shape.

However, through the same process of refinement described in the previous chapter, I've managed to see a few patterns that are sometimes present in correct answers.

Here's the list.

1. *The Three Bs: Beige, Bland, and Boring.* Because this test is standardized, and must appeal to people all around the world, the correct answer will be smoothed out. There will be no rough edges. It will be as mushy as a bowl of soggy cornflakes. Here's an quiz: Is it better to select *an unconventional type of narrative* or *the most unbelievable story about plastic sporks ever told?*

Definitely the former. The correct answer will not typically feature concrete, sensory nouns. Instead, it will use abstract nouns. Remember: *Beige, bland, and boring.*

2. *Thinking, not feeling.* The SAT is supposedly a measure of your academic abilities (though that's debatable), so choose answers that are more head, less heart. For example, if you have a choice between *excited* and *analytic,* definitely choose the latter.
3. *Grandma's House.* Do you have a nice grandmother? One whose house feels safe and warm? A place where nothing can hurt you? (If you don't, just imagine someone else's grandmother.) Anyways, picture Nice Grandmother discovering that a horrific car accident has just occurred outside her house. She shuffles over to the window, peers through the horizontal blinds. She sees bleeding people pulling themselves across her lawn, moaning, body parts everywhere, a scene of pure carnage. What does Nice Grandmother say? "Oh dear, that's unfortunate." Then she turns to you with a sweet smile and adds, "More tea, dear?" See, the lesson here is that *nothing bad enters Grandma's House.* She won't let it. She denies that misery exists. At the very least, she minimizes it nearly out of existence. That's what the bastards want from the right answer.
4. *Community, consensus, and inclusiveness.* Because the right answer is so relentlessly positive, it denies nobody. Therefore, it must accept everybody, and the result is that the right

answer very often sounds like a piece of writing that is so afraid of offending somebody that it says basically nothing at all: *When living in a foreign culture, the traveler often experiences many new opportunities.* Snooze! Remember: like most comic book movies, the right answer on the SAT must appease the widest possible group of people—and thus has virtually no identity of its own.

5. *The tone-o-meter.* Imagine a long thermometer with a plus sign at one end and a minus sign at the other end. The middle is neutral. Where are the typical answers for most questions? Slightly positive. For general tone questions, you can pretty much choose the slightly positive answer and just move on. Ninety percent of the time, you'll be right.
6. *Soft words.* Always try to pick answers with language that shows 50% possibility (as opposed to 100% certainty). Pay special attention to any answer that contains the words *can, may, might, could, sometimes, often, possible, plausible, respectful, cautious, qualified, significant, emphasize* (or *emphatic*), and especially *appreciation.* (That last one always seems to be right.)

Now let's move to the Writing questions.

WRITING

31

Question Type #7: Sentence Boundaries
Question Type #8: Form, Structure, and Sense

Once you're finished with the Reading questions, the Writing questions begin. You always know when you've crossed that border because the question changes. There is literally only one question that you will see for the rest of the exam.

Which choice completes the text so it conforms to the conventions of Standard English?

Get used to seeing this. It means the worst has passed, and you're on the home stretch of the module.

The Writing questions, as noted earlier, are easier than the Reading questions by a country mile. There is a system of rules, and you can learn them quickly. In fact, there's no reason that you can't get all the Writing questions correct.

THE FIRST TWO types of Writing questions are very similar to one another. In fact, they shouldn't even be different categories.

Sentence Boundaries questions test you on those juicy grammar nerd ideas about

- independent and dependent clauses
- punctuation
- the addition of extra information

Form, Structure, and Sense questions test you on *even more* juicy grammar nerd ideas about

- Subject-verb agreement
- Pronoun-antecedent agreement
- Verb tense
- Modifier placement
- Plural and possessive nouns

Here's an example—

MY JOB at the news desk primarily revolved around the front section and the front page. I ran pages downstairs to layout, coordinated things with the printers, distributed front-page copies to the newsroom, answered phones, and faxed the front page to our archenemy, the New York Times. They did the same to us out of professional courtesy, and the top editors always _____.

. . .

Which choice completes the text so it conforms to the conventions of Standard English?

a) will pounce on these transmissions.
b) pounce on these transmissions
c) pounced on these transmissions.
d) had pounced on these transmissions.

The answer choices are a great help in these types of questions. I like to look at the answers first to identify what's being tested. Reading vertically, you can see the only thing changing is the verb tense, so this is a Form, Structure, and Sense question.

The correct answer is D.

You know what? I feel guilty for ending that segment so abruptly. These grammar errors are easy to transmit and you would benefit a lot from them.

Can I give you a bit more specific help? Yeah?

All right, read on.

32

What follows is a list of common grammar errors that you will see on the Writing questions. Some of them are guaranteed to make an appearance on the SAT. Some may lurk instead, only popping up once a year or so.

Learn all of them!

Subject-Verb Agreement

This is the most common mistake on the grammar section. It occurs when the subject is singular and the verb is plural, or vice versa. Typically, the bastards will separate the subject and verb of a sentence and fill the space with as many prepositional phrases as possible, all of which contain nouns that are the opposite number of the subject.

Here's an example:

The petition by young people for better treatment of the whales result in ten thousand signatures.

The subject is *petition*. The verb is *result*. In between are three prepositional phrases. Look at it this way—

The petition (by young people)(for better treatment)(of the whales) result in ten thousand signatures.

Clearly, because *petition* is singular, so should be *result*.

But wait! you say. *Aren't both words singular*?

No.

Singular nouns don't carry an 's'. Singular verbs do carry an 's'.

Look—

The parrot squawks.
The parrots squawk.

Get used to finding the subject and verbs of sentences.

Singular v Plural

Closely allied with the previous error is this one. However, there are additional ways that the bastards choose to test this old chestnut. For example:

The group of students found their bags and headed for the door.

In common, ordinary spoken English, this sentence would be fine. In formal English, it's not. The reason: *The group* is singular, but *their* is plural. How to fix it?

The group of students found its bags and headed for the door.

Pretty grungy, eh? Here's another.

All those teenage girls want to become a model.

Nope, not unless they want to inhabit the exact same body.

All those teenage girls want to become models.

This is a slippery error. It's easy to overlook if you're working too fast.

Verb Tense

There are three main verb tenses: past, present, and future. There are three perfect tenses: past perfect (*had*), present perfect (*has* or *have*), and future perfect (*will have*). These six tenses are pretty much the only ones you need to know.

One common error is to use the past tense instead of the present perfect. Check this out:

Since the dawn of the twentieth century, the use of animals for transportation declined slowly.

That's wrong. You need the present perfect for this answer. Why?

Use the present perfect whenever you talk about

- unspecified past
- repeated past
- past that continues to the present (easily seen in "since" or "for" + period of time)

Since the dawn of the twentieth century, the use of animals for transportation has declined slowly.

Another common trap regards the past perfect, which describes something that occurs before the simple past. For example:

Before Joey turned seven years old, he traveled to thirty countries.

This is incorrect. *Turned* and *traveled* occurred at different times. It should look like this:

Before Joey turned seven years old, he had traveled to thirty countries.

The *travelling* occurred before the *turning*. Hence the past perfect. It's worth noting that the past perfect can only be used when there is a simple past verb in the sentence as well.

Dangling Modifier

The opening phrase of a sentence should always describe the word that immediately follows it.

Running down the street, my heel broke.

Was *my heel* running down the street? No, *I* was. You should edit the sentence so that *I* becomes the subject.

Running down the street, I broke my heel.

The dangling modifier is quite common on the exam.

Incorrect Comparison

"You can't compare apples to oranges," goes the old saying. Well, that's utterly true, especially on the SAT. In fact, if you compare something, you need to say it twice.

Here's a wrong sentence.

I like the burgers at Five Guys better than McDonalds.

What is being compared? Burgers. How many times is the word stated? Once. That makes the sentence technically wrong. This is how it should look.

I like the <u>burgers</u> at Five Guys better than the <u>burgers</u> at McDonalds.

or

I like the <u>burgers</u> at Five Guys better than <u>those</u> at McDonalds.

Using a substitute pronoun the second time—typically *this, that, these,* or *those*—is acceptable as well.

Unclear Antecedent

These can be exceptionally funny. An antecedent is a noun that a pronoun is referring to.

Jenny told Julie that she would be leaving the party.

Who is *she* referring to? It's not entirely clear. In fact, there are two ways to interpret this sentence.

Jenny told Julie that Jenny would be leaving the party.
Jenny told Julie that Julie would be leaving the party.

The first is quite likely. The second one may be rude, but it's also possible. In fact, without other context cues such as vocal inflection, hand gestures, body language, and hair-pulling catfights, we can't tell the difference.

Pronoun Misuse

Read the sentence and figure out the error.

For Terry and I, visiting Greece would be a waste of time.

Did you spot the error? If the sentence "sounded" good to you, then you're relying on a creaky, bizarre pastiche of grammar that won't do you any good on this test. Here is the correct sentence:

For Terry and me, visiting Greece would be a waste of time.

If you've ever wondered the difference between *I* and *me*—well, it's good to wonder about things. However, you do need to solidly know the difference for the SAT.

It's the difference between *subjects* and *objects*.

Subject pronouns: *I, we, you, he, she, it, they*.

Example: ___ *kissed a horse*.

Object pronouns: *me, us, you, him, her, it, them*.

Example: *A horse was kissed by* ___.

Remember that objects can follow either verbs or

prepositions. The SAT usually ignores the verbs and tests on the prepositions. For example: *at me, by me, to me, with me,* etc.

If all else fails, just follow what I call the *Rule of Thumb*: if there are two items, cover up the other one and read it to yourself. Your ear should tell you the answer. For example:

For ~~Terry and~~ I, visiting Greece would be a waste of time.

"For I"? No way. That should be "for me".

Idioms

The term idiom is very broad in English. It covers expressions, adages, aphorisms, maxims, phrases, and many other forms of lingual confusion.

On the SAT, though, it only refers to one thing: *word + preposition.*

perspective on
preoccupied with
worried about

In each of these cases, the first word means nothing without the second. They go together like peanut butter and jelly.

Sometimes, changing the preposition alone can change the meaning of the phrase:

throw up
throw out
throw down

These idioms mean *vomit, discard,* and *fight,* respectively. The SAT knows this, and it will lay traps accordingly.

Foreign students often have a lot of trouble with this, since idioms are the most difficult part of English to learn. Even students who are children of immigrants will eventually realize that they never heard idioms growing up. This is why you are urged you to read a lot.

Comparative vs. Superlative

Read the following sentence:

Of the Olsen twins, I like Ashley the best.

That's a bad sentence. Why? Because of this:

- *Comparative*: compares **two** things (*more, less, better, worse, between*)
- *Superlative*: compares **three** things or more (*most, least, best, worst, among*)

There are only two Olsen twins. Well, you could argue that they're so skinny that each one only counts for half, and therefore they count as one person.

Anyways, twins are clearly comparative. So the sentence should look like this:

Of the Olsen twins, I like Ashley <u>better</u>.

Adjective vs. Adverb

Let's review one point of basic grammar.

- An *adjective* describes a noun.
- An *adverb* describes a verb, adjective, or another adverb.

It looks like an easy distinction, but you'd be surprised how many tricky ways there are to mess this up. Here's one of them:

One interesting aspect of coastal engineering is studying the constant changing seafloor.

Did you spot the error? Here are the three words:

constant changing seafloor

Seafloor is a noun, which must be described by an adjective. *Changing* is that adjective. However, adjectives can only be described by adverbs—and *constant* isn't an adverb.

Solution: make *constant* an adverb.

One interesting aspect of coastal engineering is studying the constantly changing seafloor.

Another way they test this mistake is with that old classic, *good v. well*. Which one is the adjective? Which one is the adverb? Let's presume Tyler has just shaken off a bad case of chicken pox:

Tyler feels good.

or

Tyler feels well.

The answer is that *Tyler feels good. Good* is an adjective, and it describes Tyler at that moment. To say that *Tyler feels well* is to say that *Tyler is skilled at the task of feeling* or *Tyler has an extraordinary number of nerve endings in his fingertips.*

Comma Splice

Some students have a vague awareness of the term *comma splice,* but they usually can't describe it. If that describes you, then read this sentence very carefully:

I like cats, my sister likes dogs.

Have you written sentences like this? Two independent clauses connected by only a comma? Most teenagers have. If so, then you've committed a heinous grammar crime known as a comma splice.

There are three ways to fix such an abomination.

1. A period: *I like cats. My sister likes dogs.*
2. A semicolon: *I like cats; my sister likes dogs.*
3. A comma + FANBOYS: *I like cats, but my sister likes dogs.*

Memorize those three fixes.

By the way, FANBOYS is a mnemonic device that helps us remember the seven conjunctions in the English language:

For

And
Nor
But
Or
Yet
So

It's quite handy. Remember, if there is a transition word such as *however* or *still* between the two clauses, it needs a semicolon. Any transition word other than the FANBOYS will require a semicolon. The bastards will try to get you on that one too.

Redundant

This is an easy trick to overlook. It slides right past us when we're least expecting it. Read this:

To most people in northern latitudes, there is nothing prettier than the month of May.

Let me isolate it for you:

the month of May

Still can't see it? It's looking right at you. May is a month, so you don't need to say *the month of*.

To most people in northern latitudes, there is nothing prettier than May.

My sister was once married to a man from England. I

talked to him on the phone, asked him how he was doing. His response:

"Oh, not too good. I was made redundant last week."

I had zero idea what he was talking about. It sounded like he'd turned into a piece of grammar.

It turns out, after much awkward misunderstanding, that *made redundant* is a British idiom for *fired from your job.*

Oops.

33

Question Type #9: Transitions

THESE QUESTIONS ASK you identify the best transition for a given context.

Here's an example—

See, one dark secret about human consciousness is that we can't tolerate very much reality. Real life is messy, it doesn't feature character arcs, it doesn't have setups and payoffs, it doesn't offer dependably comedic or tragic endings—it only has bizarre randomness that leaves us scratching our heads. ____, what most of us want from our stories is internal coherence, a clear lesson.

a) In fact,
b) Thus,
c) Next,
d) Moreover,

A majority of transition words either keep the sentence going in the same direction (*therefore, consequently*) or turn the sentence in the opposite direction (*however*). A few others indicate the passage of time (*then, next*) or overcoming the objection (*nevertheless*).

To answer these questions, read everything before the blank, and read everything after the blank. Are those two ideas the same? Opposites? Change of time? Overcoming objection?

Make that determination, and simply pick the answer from the proper category.

The correct answer was B.

34

Question Type #10: Rhetorical Synthesis

THESE QUESTIONS ARE the biggest surprise of the digital SAT. It's a totally new question type.

In a rhetorical synthesis question, you're given a bulleted list of notes on a given topic. Then you're asked to identify the statement that best uses those notes to accomplish a particular goal.

Here's an example—

WHILE RESEARCHING A TOPIC, a student has taken the following notes:

- *Wars are fought over many things, most commonly resources.*
- *In the past, those resources have included precious metals, petroleum, and drugs.*

- *A new type of resource is beginning to become valuable in the 21st century: water.*
- *Access to fresh water is becoming more valuable in certain parts of the world, such as Bolivia.*
- *In these areas, conflicts over water are literally a matter of life or death, since water is so vital for life.*

The student wants to emphasize that water is more valuable than all the other resources. Which choice most effectively uses relevant information from the notes to accomplish this goal?

a) Wars have been fought for centuries over resources; today's wars are no different.
b) The 21st century has seen the continued existence of many types of precious metals and drugs.
c) In Bolivia, there is limited access to fresh water.
d) While past wars have been fought over precious metals, petroleum, and drugs, water wars are a matter of actual life and death.

Here's what to do.

Step 1: First read the question, not the notes. This will guide you in step 2.

Step 2: Read the bulleted notes. Pay attention to the ones that are related to the question.

Step 3: Go to the answer choices. Beware of answer choices that use accurate language but don't achieve the goal. And vice versa: beware answer choices that achieve the goal with inaccurate or off-topic notes.

. . .

Thus far, my students have all reported enjoying this type of question more than the others.

35

Time management tips and strategies

HERE ARE a few tips and strategies for effectively using your time on the Reading and Writing section:

- **Save time on easier questions**. Some questions on the test require more reading than others and take more time, while some question types can be answered in nearly the blink of an eye. Work swiftly on the questions that feel easier for you, and that means more time for when you need it!
- **Make predictions**. Try to come up with an answer in your mind before looking at the correct choices. Then look at the choices and select the one that best matches your prediction. This helps you avoid the silly traps that the bastards lay in every question.

- **Use the process of elimination**. On the Reading and Writing section, POE is your best friend. Use it to get closer to the right answer, especially on questions that are difficult for you.
- **Sometimes the easy answer is the right answer**. Don't worry if a question seems too obvious. If you know where the easy questions sit—usually the first one in a set of three—your instincts should guide you. The way to build those instincts is through doing practice questions and practice modules.
- **Leave time to review**. If you find yourself stuck on a question, just take your best guess and move on. Be sure to come back to the question later, once you've finished the rest of the module. How do you remember to go back? Next on the list—
- **Mark questions for review**. You can mark it right on the Bluebook app, next to the question. There's a review page at the end of the module. Click back into the questions you flagged and re-read them with a fresh set of eyes. The best way to review Reading questions is to look for evidence to support your answer in the passages.
- **Answer every question**. There's no penalty for wrong answers.

AFTERWORD

There is obviously much more to say about the SAT, but the purpose of this handbook was to provide test takers with a concise, expert way of raising their scores.

Now that you're finished, go do something else for a day or two. Then go back to the beginning, read it a second time—and take notes.

Also, let me repeat something I stated earlier.

When you're feeling down, please remember that the SAT doesn't measure anything except your ability to take the SAT. It doesn't predict future success. But it's still good to try your very best at everything in your life, because excellence is a habit. Aristotle said that over two thousand years ago. He was a smart guy, except for his clumsy attempts at science, but that's a story for a different handbook.

Good luck!

VOCABULARY LIST - 1ST TIER

Alternate forms of the word follow in parentheses.

acknowledge
anticipate
belie
benign
bombast (bombastic)
candid (candor)
capricious
compromise
contentious
controversial (controversy)
convoluted
cynical (cynicism)
digress (digression/digressive)
diplomatic
endorse
innovative
intricate
invoke (invocation)
magnanimous (magnanimity)

munificence (munificent)
nonchalant (nonchalance)
objective
partial to
partisan (partisanship)
perpetuate
political (politicized)
pragmatic
predictable
proximity
realistic
reconcile (reconciliation)
renounce (renunciation)
reticent (reticence)
revolutionary (revolutionize)
simplistic (simplicity)
subjective
toxic (toxicity)

VOCABULARY LIST - 2ND TIER

Alternate forms of the word follow in parentheses.

altruistic (altruism)
analyze (analytical/analyses)
anomalous (anomalies)
arrogance
artistic (artistry)
assuage
astute
austere
averse (aversion)
avoid (avoidance)
belligerent (belligerence)
beneficial (beneficiary)
circumspect
circumvent (circumvented)
commonplace
complacent (complacency)
complement (complementary)
complex
compunction

conciliatory
conditional
confusion
conjecture (conjectural)
contemplate (contemplative)
convention (conventional)
cosmopolitan
debilitating
delude (delusion)
dilettante
diligent (diligence)
diminutive
disingenuous
disparage (disparaging)
disregard
distort
dogmatic
dominant (dominate)
duplicity (duplicitous)
economical (economize)
edifying (edified)
eliminate
endure
enigma (enigmatic)
enthusiasm
expansion (expansive)
expedient (expediency)
expedite (expeditious)
fabricate (fabrication)
harmful
hostile (hostility)
hypothesis (hypothetical)
illuminate (illuminating)

improvise (improvisational)
indecision (indecisive)
indifference
indulged (indulgent)
ineffable
innocuous
inscrutable
insightful
inspired
irrelevant
jocular (jocularity)
laud (laudatory, lauded)
limited
loquacious
mercurial
mitigate
modest (modesty)
mollify (mollification)
monitor
novelty (novel)
obfuscate (obfuscation)
objective (objectivity)
origin (originality)
ostentatious
persist
precocious
prescriptive
proliferate (proliferation)
promote (promotion)
prophetic (prophecy)
provincial
provoke (provocative/provocation)
questionable

recognized (recognition)
redundant (redundancy)
remorseful
resilient
restrict (restrictive)
spontaneous (spontaneity)
stable (stabilizing)
subdue
succinct
superficial
superfluous
systematic
tacit (taciturn)
tenacious (tenacity)
trivial
uniform (uniformity)
variety (varied, varying)
verbose (verbosity)
versatile (versatility)
vulnerable (vulnerability)

GREEK AND LATIN WORD ROOTS

In the following list, Greek roots are indicated with (g) and Latin roots with (l). Try to learn as many as possible.

ROOT (1/g): MEANING - EXAMPLES

act (l): do - action, actor, react, transact, enact

agri (l): field - agriculture, agrarian, agronomy

alt (l): high - altitude, altimeter, alto

alter (l): other - alternate, alternative, altercation

ambul (l): walk, go - ambulance, amble, preamble

ambi or amphi (l): both, around - ambidextrous, amphibian

amo, ami (l): love - amiable, amorous, amateur, amity

ang (l): bend - angle, triangle, quadrangle, angular

anim (l): life, spirit - animate, animosity, animal, inanimate

ann, enn (l): year - annual, biennial, anniversary, annuity

anthr (g): man - anthropology, philanthropist

apt, ept (l): suitable - aptitude, inept, adept, apt

aqua (l): water- aquarium, aquatic, aquamarine

arch (g): chief - monarch, archbishop, archenemy

arch (g): primitive, ancient - archaeology, archaic, archive

art (l): skill - artisan, artist, artificial, artifact

ast (g): star - astronaut, astronomy, disaster, asterisk

aud (l): hear - audience, auditorium audiovisual, audible

auto (g): self - automatic, autobiography

belli (l): war - bellicose, antebellum, belligerent, rebellion

ben (l): good - benevolent, beneficial, benediction

biblio (g): book - bibliography, Bible

bio (g): life - biology, biography, biochemistry, biopsy

brev (l): short - abbreviation, brevity

cam (l): field - camp, campus, campaign

cand (l): glow, white - candle, candidate, incandescent

cap (l): head - cap, captain, capital, decapitate

cede, ceed (l): go, yield - proceed, exceed, succeed, concede

ceive, cept (l): take, receive - receive, reception, accept, conception

centr (l): center - central, centrifugal, egocentric, eccentric

cert (l): sure - certain, certify, ascertain, certificate

cess (l): go, yield - process, recess, access, cessation

chron (g): time - chronological, synchronize, chronicle

cide, cise (l): cut, kill - suicide, scissors, incision, insecticide

cip (l): take, receive - recipient, incipient, participate, recipe

circ (l): around, circle - circulation, circumvent

claim (l): shout - proclaim, exclaim, acclaim, clamor

clar (l): clear - clarity, declare, clarify, declaration

cline (l): lean - incline, recline, decline, inclination

clud (l): shut, closed - include, conclude, exclude, seclude

cogn (l): know - recognize, cognition, incognito

commum (l): common - community, communicate, communism

cord (l): heart - cordial, accord, concord, discord

corp (l): body - corporation, corpus, corpse, corps

cosm (g): universe - cosmos, cosmopolitan, microcosm

cred (l): believe - credible, incredible, credo

crat (g): rule - democrat, aristocrat, bureaucracy

cred (l): believe - credit, discredit, incredible

cur (l): care - cure, manicure, curable

cur (l): run - current, occur, excursion

cycl (g): circle, ring - bicycle, cycle, encyclopedia

dem (g): people - democrat, epidemic, demography

dent (l): tooth - dentist, trident, indent

dic (l): speak - dictate, predict, verdict, contradict

div (l): divide - divide, divorce, dividend

doc (l): teach - doctrine, document, doctor

don, donat (l): give - donation, donor, pardon, donate

duc (l): lead - duct, conduct, educate, induct

fac (l): make, do - factory, benefactor, facsimile

fer (l): bear, carry - transfer, ferry, infer, refer

fic (l): make, do - efficient, proficient, sufficient

fid (l): faith - fidelity, confidence, bona fide

fig (l): form - figure, configuration, disfigure

flect (l): bend - reflect, reflection, deflect

flex (l): bend - reflex, flexible

form (l): shape - form, uniform, transform, reform

fract (l): break - fracture, fraction

frag (l): break - fragment, fragile

frater (l): brother - fraternal, fraternity, fratricide

fric (l): rub - friction, dentifrice

fug (l): flee - fugitive, refugee, centrifugal, refuge

function (l): perform - function, malfunction, dysfunctional

gam (g): marriage - polygamy, monogamy, bigamy

gen (g): birth, race - generation, genocide, genealogy

geo (g): earth - geography, geometry, geology

gon (g): angle - pentagon, octagon, diagonal

grad (l): step, stage - grade, gradual, graduation

gram (g): letter, written - telegram, diagram, grammar

graph (g): write - telegraph, photograph, autograph

grat (l): pleasing - gratitude, congratulate, ungrateful

greg (l): gather - gregarious, congregation, segregation

hab, hib (l): hold - habit, habitat, prohibit, exhibit

hosp, host (l): host - hostess, hospital, hospitality

hydr (g): water - hydrogen, hydrant, hydrate

iatr (g): medical care - psychiatry, pediatrician, geriatrics

imag (l): likeness - image, imagine, imaginative, imagery

init (l): beginning - initial, initiate, initiative

integ (l): whole - integrate, integral, integrity, integer

ject (l): throw - project, inject

junct (l): join - juncture, conjunction, adjunct

jud (l): law - judge, judicial, judicious

jur (l): law - jurisdiction, jury

just (l): law - justice, just, justify

kine, cine (g): movement - kinetic, kinesthetic, cinema

lab (l): work - labor, laboratory, collaborate, elaborate

laps (l): slip - elapse, collapse, relapse

lat (l): carry - translate, relate, collate

lat (l): side - lateral, bilateral, unilateral

liber (l): free - liberty, liberal, liberate

loc (l): place - location, locate, dislocate, allocate

log (g): word - prologue, apology, dialogue, eulogy

luc (l): light - lucid, elucidate, translucent

lud (l): play - ludicrous, interlude, elude

lum (l): light - luminous, illuminate, luminescent

luna (l): moon - lunar, lunatic

lus (l): play - illusion, illusive

lust (l): shine - luster, illustrate, lackluster, illustrious

lys (g): break down - analysis, paralysis, catalyst, electrolysis

man (l): hand - manual, manufacture, manuscript, manipulate

mand (l): order - command, demand, mandate

mania (g): madness - maniac, pyromania, kleptomania

mar (l): sea - marine, maritime, submarine, mariner

mater (l): mother - maternal, maternity

matr (l): mother - matricide, matrimony, matron, matrix

max (l): great - maximum, maximize

mech (g): machine - mechanic, mechanism, mechanize

mem (l): mindful of - memory, remember, memorial, commemorate

ment (l): mind - mental, mention, demented

merge (l): dip - submerge, emerge, merge, merger

mers (l): dip - immerse, submerse

meter (g): measure - thermometer, centimeter, diameter

migr (l): change, move - migrate, immigrant, emigrate, migratory

mim (l): same - mime, mimic, pantomime

min (l): small, less - minute, mini, minor, minus, minimize

minist (l): serve - minister, administer, administration

miss (l): send - missile, dismiss, mission, remiss

mit (l): send - submit, remit, admit, transmit

mob (l): move - automobile, mobile, mobilize, mobility

mon (l): advise - admonish, premonition, admonition, monitor

morph (g): shape - metamorphosis, amorphous, anthropomorphic

mort (l): death - mortician, mortuary, mortal, immortal

mot (l): move - motion, motor, promote, demote

mov (l): move - remove, movement, unmoved

mut (l): change - mutation, mutual, commute

narr (l): tell - narrate, narrative, narrator

nat (l): born - native, nation, nativity

nav (l): ship - navy, naval, navigate

neg (l)nonegation, negative, renege

neo (g): new- neoclassic, neon, neonatal

not (l): mark - notation, notable, denote, notice

nun, noun (l): declare - announce, pronounce, denounce

nov (l): new - novel, novelty, novice, innovate

numer (l): number - numeral, enumerate, numerous

ocu (l): eye - oculist, binocular

onym (g): name - synonym, antonym, pseudonym

opt (g): eye - optician, optometrist, optic, optical

ord (l): row - order, ordinary, extraordinary

orig (l): beginning - origin, original, originate

ortho (g): straight, right - orthodontist, orthodox, orthopedist

pater (l): father - paternity, paternal, patriarch

path (g): feeling, suffer - pathology, sympathy, empathy

ped (g): child - pediatrician, encyclopedia, pedagogical

ped (l): foot - pedal, pedestrian, biped, pedestal

pel (l): drive - compel, propel, expel, repel

pend (l): hang - pendulum, suspend, append, appendix

pens (l): weigh - pensive, pension, compensate

phil (g): love - philosophy, philanthropist, philharmonic

phob (g): fear - claustrophobia, acrophobia

phon(g): sound - phonograph, symphony, telephone, phonics

photo (g): light - photograph, telephoto, photosynthesis

phys (g): nature - physical, physique, metaphysical, physician

plex (l): fold - complex, duplex, plexiglass, perplex

plic (l): fold - complicated, multiplication, duplicate

plur (l): more - plural, plurality, pluralism

pod (g): foot - podiatrist, podium, tripod

poli (g): city - metropolis, cosmopolitan, police, political

pon (l): place - opponent, exponent, proponent, postpone

pop (l): people - population, popular, pop, populace

port (l): carry - portable, transport, import, porter

pos (l): place - position, compose, deposit, composite

psych (g): mind, soul - psychology, psyche, psychiatrist

pug (l): fight - pugnacious, pugilist, repugnant

pul (l): urge - compulsory, expulsion, compulsion, repulse

put (l): think - computer, reputation, deputy, disrepute

quer, quir (l): ask, seek - query, inquiry

ques (l): ask, seek - question, inquest, request, quest

rad (l): ray, spoke - radius, radio, radiology, radium

ras (l): scrape - erase, abrasive, rasp, razor

rect (l): straight - erect, rectangle, rectify, direction, correct

reg (l): guide, rule - regal, reign, regulate, regime, regent

retro (l): back - retroactive

rid (l): laugh - ridiculous, deride, ridicule, derisive

rupt (l): break - erupt, interrupt, abrupt, rupture, bankrupt

san (l): health - sanitary, sanitarium, sane, insanity

scend (l): climb - ascend, descend, descendent, transcend

sci (l): know - science, conscience, conscious, scientific

scop (g): see - microscope, telescope, periscope

scribe (l): write - scribe, inscribe, describe, prescribe

script (l): write - script, transcript, scripture

sect (l): cut - section, dissect, intersect, sect

sed (l): settle - sedative, sediment, sedentary, sedate

sens (l): feel - sensation, sense, sensitive, sensible, sensory

sent (l): feel - consent, sentimental, dissent, assent

serv (l): save, keep - conserve, preserve, reserve, reservoir

serv (l): serve - serve, servant, service, servile

sign (l); mark - signature, signal, significant, insignia

sim (l): like - similar, simultaneous, simulate, simile

sist (l): stand - consist, resist, subsist, assist

sol (l): alone - solo, solitary, desolate, soliloquy

solv (l): loosen - dissolve, solve, solvent, resolve

son (l): sound - sonar, sonata, sonnet, unison, sonorous

soph (g): wise - philosopher, sophomore, sophisticated

spec (l): see - inspect, suspect, respect, spectator, spectacle

spir (l): breathe - respiration, inspire, spirit, perspire

sta (l): stand - station, status, stagnant, statue, stationary

stell (l): star - stellar, constellation

stimu (l): goad - stimulate, stimulus, stimulant

strict (l): draw tight - strict, restrict, constrict, stricture

struct (l): build - structure, construct, instruct, destruction

sum (l): highest - summit, summary, sum, summons

surg (l): rise - surge, insurgent, resurgent

surr (l): rise - resurrect, insurrection, resurrection

tact (l): touch - tactile, intact, contact, tact

tain (l): hold - retain, contain, detain, attain

tang(l): touch - tangible, tangent, intangible

ten (l): hold - tenacious, tenure, tenant, retentive

ten (l): stretch - tendon, tendency, tension, tent, tense

term (l): end - terminal, determine, exterminate

terr (l): land - territory, terrain, terrestrial, terrace

tex (l): weave - text, texture, textile, context

the (g): god - theology, monotheism, polytheism, atheism

therm (g): heat - thermometer, thermal, thermostat, thermos

tort (l): twist - torture, contort, retort, tort

tract (l): pull, drag - tractor, attract, subtract, traction

trib (l): give - contribute, tribute, tributary, attribute

trud (l): push - intrude, protrude, intruder

trus (l): push - intrusive, obtrusive

turb (l): confusion - disturb, turbulent, perturb, turbine

urb (l): city - urban, suburb, urbane, suburban

vac (l): empty - vacant, vacation, vacuum, evacuate

vag (l): wander - vagrant, vague, vagabond, vagary

var (l): different - vary, invariable, variant, variety

ven (l): come - convene, convention, advent, invent

ver (l): truth - verdict, verify, veracity

ver (l): turn - convert, reverse, versatile, introvert

vict (l): conquer - victory, victim, conviction

vid (l): see - video, evidence, provide, providence

vinc (l): conquer - convince, invincible

voc (l): voice - vocal, advocate, convocation

void (l): empty - devoid, avoid, void, voided

vol (l): wish, will - volition, volunteer, voluntary, benevolent

volv (l): roll - revolve, involve, evolve, revolution

vor (l): eat - carnivorous, voracious, herbivorous

PLOTWORKS PUBLISHING

for dinner in a formal black gown at the college dining hall—

From punting on the Cherwell River to sitting for a debate at the Oxford Union—

From crowded parties in Scotland to lonely forests in Wales—

The Oxford Diaries is a romantic snapshot of an undergraduate experience that only comes along once in a lifetime.

Be transported to a world of literature, history, fantasy, and tradition. Read *The Oxford Diaries* today!

THE
OXFORD
DIARIES
A STUDENT TRAVELOGUE
J.A. JERNAY

THE OXFORD DIARIES

September 5

In lecture today, Dr. Cox-Davies touted the virtues of Oxford. This city, he said, is the sixth-largest in England, and it was a center of commerce long before it was a college town. Two of the three roads in ancient Saxony passed by either side of it.

He told a funny story about the clash between town and gown. Back in the thirteenth century, the townspeople in Oxford frowned upon the scholars for their laziness. Easy to understand why, since the medieval scholastics were engaging in prolonged debates about how many angels could dance on the head of a pin.

Things reached the breaking point when a scholar decided to practice archery from his bedroom window. He took aim at a basket of apples on the head of a townswoman passing by. He missed and accidentally killed the woman. Enraged, the townspeople rose up and slaughtered half the scholars. Sounds like a drunken dare gone horribly wrong.

The remaining scholars had a choice to make. They

could stay in Oxford, where they were hated, or establish a new school, far from the nosy intrusions of the urban populace. A portion of the scholars elected to leave, and they traipsed across the fenlands, where they founded Cambridge University, out in silence and peace.

"And that is why," he said, "Cambridge is on the way to *nowhere*."

I totally love silly rivalries, and the more ancient, the better.

I owe a debt of gratitude to Dr. Finlay. Today, with the weight of his name and with the stern, headmistress tone of Rosie Robertson—the Registrar with a Ramrod up her Rear —I finally have a bank account at Westminster Bank. The bank hadn't wanted to allow a nineteen-year-old from the U.S. to open an account, but the college personally vouched for me. Whew. Now I can buy food again.

I took an oath at the Bodleian Library today. I promised not to "kindle any flames" in the stacks, nor to remove a book from its reading room. My question is this: Why am I swearing fealty to a dead man whose scions won't even allow me to *use* the damn library after October 7, when all the British students—aka "real" students—hit town? Would they have extended my privileges if I had recited the oath in Latin?

The entire Oxford University library system is just as frustrating. It's nearly useless because of inefficiency. The Bodleian has apparently been long abandoned by all *real* scholars who harbor any notions of speedy research. The library at my satellite college is only open for two hours a day, and books take an hour to be removed by somebody who is standing only thirty feet away from them. I'm even told that the Christ Church library still chains its books to the shelves.

Made in the USA
Monee, IL
06 November 2026

68012408R00090